# Organic Farming

## *How to Start and Maintain Your Own Organic Farm*

*David Sykes*

## Table of Contents

Introduction

Chapter 1: Organic Farming

Chapter 2: Going Organic

Chapter 3: Weed management

Chapter 4: Insect Management

Chapter 5: Tillage

Chapter 6: Soul of the Soil

Chapter 7: Vegetable Plant Guide

Chapter 8: Quick Tips and Reminders

Conclusion

# Check Out My Other Books

# Introduction

I want to thank you and congratulate you for purchasing the book, *"Organic Farming"*.

This book contains proven steps and strategies on how to get started with Organic Farming. This will also serve as your guide in maintaining a healthy organic farm. Here, you'll find helpful tips that are easy enough to apply firsthand. There is also a special section on growing vegetables. It will tell you exactly how to plant and grow vegetables organically. Most importantly, this book will teach you how to keep your organic farm for many, many years.

Thanks again for purchasing this book, I hope you enjoy it!

© **Copyright 2015 by David Sykes - All rights reserved.**

This document is geared towards providing exact and reliable information in regards to the topic and issue covered. The publication is sold with the idea that the publisher is not required to render accounting, officially permitted, or otherwise, qualified services. If advice is necessary, legal or professional, a practiced individual in the profession should be ordered.

- From a Declaration of Principles which was accepted and approved equally by a Committee of the American Bar Association and a Committee of Publishers and Associations.

In no way is it legal to reproduce, duplicate, or transmit any part of this document in either electronic means or in printed format. Recording of this publication is strictly prohibited and any storage of this document is not allowed unless with written permission from the publisher. All rights reserved.

The information provided herein is stated to be truthful and consistent, in that any liability, in terms of inattention or otherwise, by any usage or abuse of any policies, processes, or directions contained within is the solitary and utter responsibility of the recipient reader. Under no circumstances will any legal responsibility or blame be held against the publisher for any reparation, damages, or monetary loss due to the information herein, either directly or indirectly.

Respective authors own all copyrights not held by the publisher.

The information herein is offered for informational purposes solely, and is universal as so. The presentation of the information is without contract or any type of guarantee assurance.

The trademarks that are used are without any consent, and the publication of the trademark is without permission or backing by the trademark owner. All trademarks and brands within this book are for clarifying purposes only and are the owned by the owners themselves, not affiliated with this document.

# Chapter 1: Organic Farming

You may not know it, but you have probably eaten fruits and vegetables that were grown organically. But what is organic? What does it do to our body? And how do you actually grow food organically?

Organic is a term used for a variety of food that is produced through methods approved by the National Organics Program and the US Department of Agriculture. Being organic is about not using any form of chemical in growing fruits and vegetables. It is the best way to go as it promotes good health. Organic food tastes a lot better than those conventionally-grown. The only downside is it can be a little expensive to keep buying food grown organically. It is better to have your own organic farm so you can save a lot of money and earn at the same time.

Starting an organic farm has many benefits. Apart from earning money from the produce you grow, you become healthier by consuming organic food. However, there's more to organic farming than just planting seeds without using pesticides and other chemicals. It's a system specifically designed to help the environment and the people living in it. Although it takes a lot of time and effort to grow and maintain an organic farm, it's not difficult so long as you adhere to the rules and regulations set by the US Department of Agriculture. Don't worry, this eBook will teach exactly what to do with your organic farm. Just keep reading for helpful tips and guides on how to grow and maintain an organic farm.

*Why farm organically?*

It is always better to grow your own food. That way you would know how and where your food came from. And apart from the many benefits you'll get when you grow your food organically, you are also helping other people consume healthy produce.

*Why should a consumer buy organic?*

There are many reasons why consumers purchase organic food. Most people simply want to use products that are free from chemicals which could potentially harm them. There are also consumers who just like to try new organic products. Others want better-tasting produce. It has been proven that organically grown food tastes a lot better than those produced by conventional farmers.

Organic products are expensive because it takes time to harvest them. At the least, organic produce will be harvest in a year. Organic farmers need to exercise extra care for their plants to ensure proper growth. And of course, they need to be more patient too.

*The transition to Organic Farming*

The first few years would be the hardest. There are rules that need to be adhered to when you decide to switch to organic farming. Although it will be rewarding, you need to still follow and adjust to the system. Each organic farmer is required to manage their lands and submit their organic plans to the US Department of Agriculture. It will be reviewed and checked whether farmers are really growing their food organically – without any chemicals. Reviewing and checking would entail visiting the farms and directly checking on each plant.

Organic farmers need to be really consistent and religious in following rules and strategies. The transition period is 36 months prior to harvest of the first organic grown crop. Cash flow will also pose as a problem during the transition period. This is because the organic farm is still unstable and the products still will not be qualified as a certified organic. What the farmer can do is grow some low-cost crops during the transition period to not incur a lot of expenses. He might want to also start his marketing early so he can sell his crops immediately.

You need to carefully plan everything if you want to be successful as an organic farmer. It may take up to 10 years to become completely organic. But don't worry, your hard work will all pay off. The trick is to remember the number one rule: No using of chemicals in any form.

## Chapter 2: Going Organic

Starting an organic farm isn't as complicated as you think. Farmers who are used to growing produce conventionally may find it a little hard to switch to organic, but it's achievable through hard work and patience. You will need to completely change how you grow your own food. Don't worry as this is for your own benefit. I'm pretty sure you're already aware what organic food can do to your overall health.

In organic farming, you can't use:

- Pesticides
- Synthetic fertilizers
- Genetic engineering
- Growth hormones

Follow these guidelines in growing your own crops:

The US Department of Agriculture has a website specifically for organic farmers. They are actually provided with the resources they need. Farmers need to just visit the website and keep themselves informed and updated with how everything works in organic farming. It's to avoid future problems with their organic farms.

Check the guidelines for the prohibited substances that should not be applied to the land at least three years before the crops are harvested.

Here's another important thing to remember: The organic farm should not be near any land that's using substances that are prohibited.

*What is USDA certification?*

The only way for an organic farmer to get a USDA Certification is when he adheres to the rules all the time. There are no exceptions. Everyone must follow else they will never be certified as an organic farmer.

A certification is a process that involves checking whether the product being marketed and sold is 100% organic. This is based on the requirements set by The Department of Agriculture. All producers of food labeled as organic should get the certification. It will serve as proof that what you are selling is indeed organic.

It's not difficult to get certified, but it will take a considerable amount of time.

*Benefits of Organic Farming*

As opposed to conventional farming, going organic has many benefits. Here is a detailed list as to how Organic food improves overall health.

- **Better Nutrition** - The nutritional value of food is based on its vitamin and mineral content. Compared to food grown conventionally, organic food provides the best nutrients. Organic farming also gives out nutrients to the soil which is then passed on to the plants and the animals.

- **Better Taste** – Organically-grown food taste a lot better than food grown with a lot of chemicals. The sugar content in organically grown fruits also tastes best.

- **Longer Shelf-life** – Organically-grown food last longer than those not organically grown.

- **Lower Costs** - Contrary to what people believe in, organic farming isn't costly. As a matter of fact, you would be able to reduce costs when you grow your own organic food. With conventional farming, there is a need to buy chemicals worth thousands of dollars just to make the produce grow. Organic farmers do not need to spend more money in growing their food. It would only require a lot of patience and extra care for their plants.

- **Environment-friendly** - Chemicals contaminate the soil and other sources, which can negatively affect plant life. Also, we are affected by these chemicals. With organic farming, the environment doesn't suffer. Moreover, we do not suffer as well.

There are disadvantages to organic farming too, but the pros far outweigh the cons.

*Steps to growing an organic farm*

The first step to growing an organic farm is to get a farm lot and decide which crops you plan to grow on your farm. Growing an organic farm would also entail knowing everything you can about cultivating soil and keeping pests away. Avoid synthetic poisons when keeping pests away. When you buy crops, make sure the seeds are also organically grown, else your produce will not be certified organic.

### *Prepare the soil*

A healthy soil helps provide better nutrients to the plants. Better nutrients means better health. Never treat your soil with chemicals if you want to grow an organic garden. Chemicals can harm you and your plants.

The first thing you need to do is get your soil tested. You may get a home testing kit or send a sample directly to a local agricultural office. When you get your soil tested, you would know the breakdown of the nutrient levels of the soil. You will also get recommendations from the experts on how to better treat your soil without chemicals.

### *Good compost*

Any garden can benefit from compost. And you can make your own. One benefit of compost is it will keep your crops healthy.

Here's how to make your own compost:

- Get at least three feet of square space in your garden for your compost. Add leaves and garden trimmings with layers of soil in between. Apply them alternately.

- Use natural fertilizers. This is crucial. Using natural fertilizers can make the soil healthier. You get better fruits and vegetables with natural fertilizers. Add a little water to keep the compost moist. Top it off with up to six inches of soil.

### *Buy the right plants*

This is an essential step. Make sure the plants or crops you grow in your organic garden are the right ones. You need to choose the right plants that will grow in your soil. You also need to consider your climate. If it's the cooler months, do not buy crops that don't grow in the cold season. If you don't know which crops grow in the hotter or cooler months, you may want to check with The US Department of Agriculture or your fellow farmers. For sure they will be glad to assist you.

Water them properly. Plants need water to survive. It is essential that you know when to water them to keep them alive and healthy. Mornings are the best time to water your plants. Avoid watering your plants at night as it is more likely to be damaged by a lot of bacteria.

Protect your plants. To avoid pests, make sure your plants get proper sunlight, moisture, and nutrients. Plants need at least six hours of full sun.

## *Organic Livestock*

There are standards that need to be followed when raising organic livestock. Verification agencies will check whether an organic farmer is indeed following the set standards for livestock farming. Below are the key considerations in livestock farming.

- Livestock feed
  All the rations of the livestock must also be organically produced under the same sets of standards.

- Waste Management
  To avoid contamination of crops, organic farmers are encouraged to manage animal manure.

- Origins of the livestock
  All livestock that are raised and sold must also be raised organically under the same sets of standards by the US Department of Agriculture.

- Record keeping

Records are essential for tracking purposes. It's also used to verify some information needed.

# Chapter 3: Weed management

Weeds are unwanted plants in the garden. If not properly managed, they can damage crops a lot worse than other garden pests. Controlling them is one of the biggest challenges in organic farming.

Just in 2009 alone, the Food and Agriculture Organization of the United Stations estimated a total loss of $95 billion due to the damaging effects of weeds. This is, in part, because there are organic farmers who do not yet know how to do weed management effectively.

The first step to weed control is becoming aware of the types of weeds that will grow on your farm. That way you'd know how to effectively deal with them. A long term plan should be taken into consideration when controlling weeds. It is easy to get rid of them in conventional farming, but the use of chemicals is not allowed in organic farms.

These are the types of weed species that you should look out for in your organic farm:

*Annual species*

One of the most effective ways to get rid of annual species is to control their growth in the early season. That is before they actually start competing with other crops. Although they will die within a year after germination, it may be difficult to control them if they begin to develop. And also, you would already have done a lot of damage to your crops. You'd lose more money if you don't get rid of annual species right away.

*Biennial species*

This type of species lasts for two years. That's really a long time. Soil cultivation is the best way to get rid of biennial species. So before they begin to flower, make sure you already know how to effectively cultivate your soil to get rid of biennial species.

*Perennial species*

They are more difficult to control compared to the first two species. You might want to directly get rid of them by ploughing.

*Creeping perennials*

This type spreads by the roots. It's not a pleasant sight at all, and it can damage a lot of crops.

*Stationary perennials*

This is the type that will cause trouble in arable crops and pasture species. They occupy a large area which could be used for other crops. If you can't manage this type of weed, it'd do great damage to your organic farm. And you will lose money too.

There are different methods for controlling weeds. For organic farming, no chemicals in whatever type should be used. Organic farmers need to use organic methods instead. They include:

**Cultural methods**

This method includes crop rotation, crop establishment, and soil cultivation.

*Crop rotation*

This method is essential in managing weed growth. It's better to have different variations of crops to prevent certain weed species to grow abundantly.

*Cover Crops*

You would be able to suppress weeds with cover crops. Also, they have other benefits. They include: improvement of soil structure and enhancing soil fertility, protection against soil erosion, reducing population of weeds and weeds seeds.

**Direct-control methods**

This method includes biological control which means the farmer should use insects, bacteria, and other animals in getting rid of the weeds.

You may want to also look at this list of physical methods of controlling weeds:

*Plough*

What it does: Ploughing works best with perennial weeds. It can prevent seeds from growing. They can also bury seeds that have already germinated.

*Harrow*

What it does: Although it may stimulate seed germination, harrowing can really kill and destroy weed plants that can do a lot of damage to your crops.

*Weed harrow*

What it does: weed harrowing can cover small weed plants with soil. It can also uproot them.

*Cultivator*

What it does: This can actually disrupt the growth of weeds and can help prevent the production of seeds. One good thing about a cultivator is it can bury seeds that were produced in the same year.

*Brush weeder*

They can cover small weeds with soil. They can also uproot them.

*Weed mower*

They can cut weeds that can damage crops.

What to do during climate change?

Unfortunately, weeds can adapt better with climate change. That's why it is important for an organic farmer to devise crop-breeding programs to effectively manage the growth of unwanted plants.

For an organic farmer, getting rid of or eliminating weeds is never easy. While a conventional farmer can rely on chemicals in controlling weeds, the organic farmer needs to think of a long-term solution for it. He needs to be aware of the different types of weeds for him to be able to effectively manage or control weeds.

Crop rotation can suppress population of weeds. It is actually essential in organic farming. It's also better to combine different kinds of weed control methods to be able to get rid of them right away.

# Chapter 4: Insect Management

In organic farming, the idea is not to eliminate insects but to manage them. An organic farmer can successfully manage insects if he knows what they need to survive, how they interact with the environment, and how they can be manipulated to protect your farm. As soon as you find out more about insects, you would be able to draft a plan and incorporate different strategies in effectively managing them to protect your crops.

Getting rid of pests is important in organic farming. Pests can also damage crops and will make you lose money. Here are some things you can do to get rid of pests without using any chemicals:

*Crop rotation*

Rotating crops can promote soil fertility. It can also help in getting rid of pests that can damage your crops. Crop rotation involves altering the type of crops that a farmer grows on his organic farm. When they alternate crops, the species will not get used to the type of plant that is being cultivated. Crop rotation is a lot better than using chemicals in getting rid of pests. If you have an organic farm, using chemicals is a no-no. Chemicals can damage the crops as well as the soil that is being cultivated.

*Intercropping*

This method will make it harder for pests to target a particular crop. Intercropping involves simultaneous cultivation of more crops but on the same field. When you plant different varieties of crops on the same field, the distance between types increases, making it difficult for pests to stay on a particular crop or target a main crop.

*Use other pests in getting rid of pests*

Predatory mites and ladybugs can actually get rid of pests. Knowing how to manage insects will help you get rid of other pests.

# Chapter 5: Tillage

Farming can be a really tedious task. It's not easy to prepare your farm for growing different types of crops. There are things you needed to do like checking your soil, and choosing the right crops to grow on your farm. If you have a bigger farm, you might consider a tillage to help you prepare your land for growing your crops

*What is a Tillage and how it benefits organic farmers*

Tillage is a process that helps a farmer prepare his land for planting different varieties of crops. Although it is not advised to really depend on tillage practices for weed control, it is useful in preparing the soil for planting.

Tillage is a modification of the structure of the soil. It involves milling, beating, crushing, and cutting. There are advantages and disadvantages to using tillage for organic farming. Experienced farmers can reduce the disadvantages of tillage by carefully planning how to use them.

As part of their organic-system plans, farmers are asked to document their tillage procedures.

Some benefits of tillage:

*It will condition the soil*

This is needed to grow healthy crops. Soil conditioning also favors agronomic processes.

*Mixing and incorporating farming needs*

Manures, seeds, and fertilizers are needed in the organic farm. Tillage can help out in ensuring that some of these substances are evenly mixed and distributed.

*Land forming*

It will prepare your land for planting crops.

*It can suppress pests and weeds.*

It can directly kill weeds and get rid of pests that can harm your crops.

Tillage also poses some disadvantages, so before considering using a tillage operation in your organic garden, you need to be really clear as to the purpose of using tillage practices. Some tips you can use:

- Ask yourself if you really need tillage. Otherwise, use other methods.

- Check your soil before considering using tillage. If it's too wet or too dry, avoid tillage.

- You do not have to cultivate deeply to control the weeds.

Tillage practices are beneficial, but you do not really need to rely on them all the time. Always weigh the pros and cons to make a sound decision.

# Chapter 6: Soul of the Soil

The soil matters when farming. A good soil enables plants to grow and bear fruit. A bad soil will not give you good and healthy crops, or it won't really let you grow anything.

How do you improve garden soil to ensure crop growth? Follow these tips:

You can use natural methods like adding a little compost for your vegetables, but in the long run, you need to cultivate your garden soil better for your garden to produce high quality crops. Here is what you can do:

*Add manure*

You can use livestock manure to keep a good soil. In fact, it contributes a lot to soil aggregation. Just one important thing to remember: You should allow up to three months between applying manure and harvesting some of your crops like green leafy vegetables. You don't want to contaminate your crops with livestock manure.

*Go for composting*

Composting is the recycling of any organic wastes. It adds more nutrients to the soil which will then go to the plants or crops. Try to apply appropriate amounts of compost per season and watch your soil improve.

If you use the old composting method, you need to build taller piles of bins then alternate from putting in greens, grass clipping, and dry leaves. Another alternative is vermicomposting. This is when you use earthworms, manure, green crop residues, and food wastes.

*Use cover crops*

This is one of the best strategies that you can use to feed your soil and enhance its structure eventually.

*Green Manure Crops to heal soil*

The best thing about green manure, apart from healing soil, is it can help eliminate weeds without using chemicals. Green manure is also a lot more affordable than animal manure.

**Here are 4 crops that will provide more nutrients to your soil:**

Red Clover. This crop has amazing benefits not just to humans but to your soil as well. It will help revitalize it. When planting red clover, make sure the seeds are at least 6 inches apart.

Winter tare. This will help prevent soil erosion. This type of crop can stand the winter season. It could grow up to 12 feet, so it can actually act as protection for the soil when it gets too cold.

Lupin. This is a legume. You can eat Lupin or use it to heal your soil.

Broad Beans. This crop has a lot of nutrients that your soil needs. Make sure you plant one of these for your organic farm.

### *Maintaining soil fertility*

A healthy soil also produces healthy plants. Here are some tips to help you maintain the fertility of your soil:

Mulch as much as you can. Mulching is the process of covering the surface of the soil with plant residues to keep its moisture. It will prevent soil erosion and improve soil structure. Mulching can also add nutrients to the soil and it can suppress weeds.

Dig less. Dig deeply for a start. Try to remove hard layers in the soil before applying compost. As much as possible, do not disturb the soil for a couple of years. Just apply the compost in the surface. Stop digging deeply.

Always apply compost. Applying compost is important when farming. Compost is the life of your plants. Without it, they won't grow. It also helps in maintaining the structure of the soil as well as prevent soil erosion.

# Chapter 7: Vegetable Plant Guide

*Asparagus*

The best time to grow asparagus is during the spring. It's not easy to grow this vegetable organically, but if you're patient enough to learn how to grow it, an asparagus can provide you with a lot of benefits. You need to have good soil to be successful in growing an asparagus in your organic garden. Learn how to cultivate soil better to be successful in planting an asparagus.

Some facts you need to know:

It needs at least 6 hours of full sun.

An asparagus can survive cold climates.

If you want to get the maximum benefit of this vegetable, try planting this on a separate spot. Some companions of asparagus you may also consider planting: parsley, basil, tomato, and carrots.

Since an asparagus is a perennial plant, it will take at least a year for it to be harvested. However, when it starts growing, it will keep on producing asparagus for up to 15 years or more. That is why it is important to have good soil to encourage healthy growth of this plant.

If you want to harvest an asparagus quickly, try to buy a year-old asparagus crowns. Asparagus seeds take longer to grow.

## *Cabbage*

Planting cabbage is easy for beginners. It is also easy to store, so you can be assured that cabbages will be with you until winter time.

When planting cabbages, make sure they get plenty of sun and good soil. Since it is an annual plant, it can be harvested 70 days after planting. Cabbages can also stand cold climates. They need up to 6 hours of full sun.

Some things you need to know:

Planting cabbage seeds should be done indoors first, about eight weeks before planting them in your organic garden. Get a good soil and plant them about one quarter inch deep. Avoid planting them too deeply as they won't grow.

The temperature of the soil should be at 72 degrees minimum for up to five days. And then, drop it back to 55 degrees. This is to prevent the seedlings to becoming leggy and thin.

You may harvest the cabbages as soon as the heads of the plants have already formed. Just cut them when they are already firm else they will crack.

## *Carrots*

Although it would require a lot of patience, growing carrots is easy. This is also an annual plant so you would be able to harvest them 70 days after planting. Recommended variety of carrots that should be planted are sugarsnax 54, red cored or oxheart. They also need up to 6 hours of full sun.

Some things you need to know:

Make sure the soil has a lot of nutrients as they grow best in it. When planting carrots, make sure the seeds are about 12 mm apart. The carrot seeds prefer warm and moist soil, so what you can do is place plastic over them, especially if you live in a colder climate.

If the carrot seeds are planted thickly, thin them as soon as they are 3 inches tall. Just avoid excessive thinning to avoid carrot root fly. The temperature should not be below 70 degrees for ideal taste.

When harvesting carrots, pull them by hand. Don't harvest carrots when it's too hot.

## **Lettuce**

It is better to plant lettuce early in the year as they mostly thrive in cooler weather. Choose the type of lettuce that will continue to keep growing for months after harvesting. Lettuces need at least 6 hours of full sun. If you plan to plant lettuces, consider planting their companion plants such as strawberry, carrots, beans, cucumber, and radish. Avoid planting broccoli, grains, and fava beans with lettuces.

Some things you need to know:

Leaf lettuces are best grown in cool weather. They are easy to grow from seed. The head variety, however, can tolerate heat more than the leafy varieties. One important thing to note: Whatever variety of lettuce you choose, make sure the seeds get enough sunlight. They should have at least six hours of sun exposure.

Plant the lettuce seeds about 0.2 inches deep. They will germinate 6-12 days after planting.

Lettuces are better planted in cooler weather, but if you want to grow lettuces when it's warmer, make sure you provide a shade for them. They will become temporarily dormant when it gets too warm.

### **Tomatoes**

Growing tomatoes is easy, but you need to maintain warmer surroundings for them to grow.

Some tips for growing tomatoes:

Tomatoes are an annual plant so they can be harvested 85 days after planting. Tomatoes cannot survive in cold soil. Make sure the soil is at least 32 degrees Fahrenheit. They also need up to 8 hours of full sun.

When planting tomatoes, make sure they are 2.5 cm apart. Plant them about 1/8 inch deep. They usually germinate in 6- 10 days.

If you want more tomatoes, make sure you only use an organic fertilizer. Spray it onto the tomato plants several times to get its full benefit.

If you live in colder places, cover the soil to keep them warm. When watering, wait for the soil to dry out.

When you follow the guide above, you should be able to grow these crops the right way.

# Chapter 8: Quick Tips and Reminders

At this point, you should already be quite knowledgeable about organic farming. Still, given that it's your first time engaging in such an endeavor, I've prepared this list of reminders and tips – just to make it easier for you to check a few things once you've already started to get your hands dirty.

- It is vital to keep monitoring the plants or crops closely to be able to accurately understand their nature and any problems that may arise.

- When dealing with weeds, always consider the physical measures (like using traps, handpicking, etc.) and creating protective structures (like shades and greenhouses).

- Always remember that not all insects are pests. Ladybirds, spiders, and wasps help eliminate pests. You will be able to reduce costs when you use insects to fight pests.

- Never use pesticides or any chemicals when dealing with problems in your organic farm.

- Sterilize all farm tools immediately.

- Weeds and pests pose a major problem in an organic farm. Make sure you know all the possible measures of controlling them without using any chemicals.

- Always water your plants in the morning.

- Always ask for help from other farmers. Don't be afraid to ask. If there are other farmers near your area, contact them if there are things you are doubtful about. Just make the sure the farmers are also trying to maintain organic farms.

- It is better to check your land first, soil, and the condition of the climate before deciding which crop to grow in your organic farm.

- Keeping the soil fertile and ensuring proper nutrient management should be among your priorities. Always improve the soil's condition and work on preventing erosion.

# Conclusion

Thank you again for purchasing this book!

I hope this book was able to help you start your own organic farm.

The next step is to make sure you religiously follow the guidelines provided on this book.

Finally, if you enjoyed this book, then I'd like to ask you for a favor, would you be kind enough to leave a review for this book on Amazon? It'd be greatly appreciated!

Click here to leave a review for this book on Amazon!


Thank you and good luck!

# Check Out My Other Books

Below you will find some of my other popular books that are popular on Amazon and Kindle as well. Simply click on the links below to check them out. Alternatively, you can visit my author page on Amazon to see other work done by me. If the links do not work, for whatever reason, you can simply search for these titles on the Amazon website to find them.

**1) The Miracle of Garlic: Herbal Remedy for Weight Loss, Diabetes, Blood Pressure, Cholesterol, Cancer, Allergies and Much Much More**

go to: http://amzn.to/1UxOGXp

**2) The Miracle Of Green Tea: Herbal Remedy for Weight Loss, Diabetes, Blood Pressure, Cholesterol, Cancer, Allergies and Much, Much More**

go to: http://amzn.to/1PoITFT

www.ingramcontent.com/pod-product-compliance
Lightning Source LLC
Chambersburg PA
CBHW030517220526
45464CB00006B/2840